CREATION and SCIENCE
RECONCILED

Archy T. Gault

To my dear friends
James and Angela
with sincere affection

Archy

Pen Press Publishers Ltd

© Archy T. Gault 2004

All rights reserved. No part of this publication
may be reproduced, stored in a retrieval system,
or transmitted in any form or by any means,
electronic, mechanical, photocopying, recording,
or otherwise, without the prior permision
of the publishers.

First published in Great Britain
Pen Press Publishers Ltd
39-41 North Road
London N7 9DP

ISBN 1 904018 75 0

A catalogue record for this book is available
from the British Library

Printed and bound in Great Britain

Artwork by James McMichael
of Ballycastle

Cover design Jacqueline Abromeit

Dedication

This study is dedicated to my wife, Janette,
without whose support, patience, understanding
and encouragement this study would not
have been researched and completed.

Acknowledgements

In the first instance, these go posthumously to my Father and Mother, John and Anna Gault, who gave an upbringing in which I was... *'Rooted in Him, and established in the faith, as...taught'* (Col. 2/7). This grounding led me to further study of and exploration of the Scriptures.

Next I must express my sincere appreciation to my late friend, Alex Aird, for his willingness to be my sounding board, during the early stages of research and development of the study.
His constructive criticism and encouragement were welcomed.

I must now thank my wife's sister, Mrs Rena Cunningham, whose comment that my study, which had previously been about Time from its creation to its cessation, could be more easily understood if divided into parts.
This first part, being from the beginning of Time to the beginning of recorded Biblical time.
The other part forms a separate study.

James McMichael is to be congratulated on his inspired artwork, part of which forms the basis of the book cover design. He is also to be thanked for his honest written critique, which is an important part of the completed study.

My cousin Rosemary Dobbin's assistance in the proof reading, is very much appreciated, as was her verbal critique of the various drafts, during the stages leading up to the preparation of the final typescript.

Archy T Gault

Contents

Author's Note

Preface   1

Chapter 1
    Original Creation   9

Chapter 2
    Angelic Rule   25

Chapter 3
    Void and Dark   31

Chapter 4
    The Replenishment of the Earth   35

Chapter 5
    The Creation of Adam   49

Independent Questions   63
Summary   65

AUTHOR'S NOTE

Now that this Study is complete and being published, I find myself questioning my reasons for undertaking the work involved. Whether this is usual for authors or not, I am unsure, but feel that if I were to answer the various questions which were in my mind, it may assist others in the understanding of my motive for writing it, and also possibly follow the content of the Study more easily.

In general terms the Study has been based on the Scriptures, and the King James Version is that used and quoted, with reference being made to the editions of Dr. C I Scofield and Thomas Newberry, where the marginal references give a more accurate translation of the meaning of various words. A simple example of this is in Genesis 1/6, where the word 'firmament' is used, and if this word is taken as seen or sounding, it gives the impression of something of a solid nature. Both Scofield and Newberry's marginal references indicate that the true translation is of ***"an expanse between the waters beneath, of vapour above"*** and this gives a clearer understanding of what had been created.

It must be emphasised that the Scriptures used are quoted as written and include no personal interpretation of mine. Although Biblically based, the Study should not be interpreted as being a Christian evangelical message, it being purely an historical study into the Creation by Almighty God and this Creation's reconciliation to scientific facts.

I feel that it should be noted here that the Scriptures have in the main four particular elements, these being those of historic, prophetic, evangelical and doctrinal importance, but in this Study I have concentrated on the historic element alone.

In doing this I feel that it may be of interest to those who may be agnostic, and even atheistic, but who do have a real interest in the subject matter studied.

It is also known that there are many people who believe in God, but are not affiliated to any religious organization, and who also may find it of some interest.

The Study does have a Christian basis and will, hopefully, create discussion and produce some positive criticism.

It could have interest for those of the Israeli/Jewish faith as the Scriptures used contain the Pentateuch, (the first five books of the Bible) which is the will of God as revealed in the Mosaic Law, and is known in the Israeli/Jewish faith as the Torah. They also include the Talmud, which is the body of Jewish civil and ceremonial law, comprising the Mishnah and Gemara.

At this point it should be noted that "...[the] faith..." referred to in Jude v 3 is given a definition in Hebrews 11/1 which is supportive of my comments above, in that there are people mentioned in later parts of Chapter 11, such as Abel, Enoch, Noah, Abraham and Sarah, and many others who all *"...died in <u>faith</u> not having received the promise..."*

The brackets around the word 'the' indicate that this word was not in the original document.

The *faith* referred to in Jude can be seen to have had significance for many people who lived from Adam to Christ, and it can therefore be deemed to have been a universal faith, and continues to be so.

Those who lived from Adam to Moses were not affiliated to any recognized religion, but lived by faith in God, whilst those from Moses to Christ belonged to the Israeli faith.

This Israeli faith was established as a religion when God spoke to Moses saying, **"This month shall be unto you the beginning of months…" (Exodus 12/2).**

It later became the religion of the nations of the 10 tribes of Israel under Jeroboam, and of the tribes of Judah and Benjamin under Rehoboam. (1st Kings 12/16-24).

This latter nation of 2 tribes becoming in time, the Jewish Nation.

Therefore, it being a God based universal faith, it can be taken as being unchanged from Adam to Christ.

The reason for undertaking this Study is mainly subjective, as my mind was in a state of dichotomy, in that my upbringing gave me a true belief and faith in God, which not only applied to His existence but also to the acceptance of all of His works, whether I could understand them or not, and I still hold firmly to this position.

That the Study is also objective can be seen from the previous paragraph.

The dichotomy arose in my mind when, in studying the History of Architecture as part of my professional training, I came across buildings and civilizations which predated the Biblical date of 4004BC arrived at by Archbishop James Ussher. This date was generally accepted as being correct, and will be shown in this Study as being incorrect, and should be 6274 BC. The reasons for this correction are given in the text of the Study.

One of these civilizations was that of Mohenjo Daro, which was discovered in the sub-continent of India, and which apparently existed many years prior to the date arrived at by Archbishop James Ussher.

The existence of similar civilizations was proven in other Continents, all of which caused me to try to resolve this dichotomy.

These all related to the time factor differential between accepted Biblical teaching and the continuing discoveries of proven scientific facts.

These scientific facts have to be accepted, and upon acceptance, the reconciliation of these scientific facts to Biblical truths is possible.

I realized that the answer possibly lay in the Scriptures, and that an exploration and deeper study of these was necessary to enable this reconciliation to take place. I knew that the answer was there if the Scriptures were taken as written, and not in Man's interpretation of them, and indeed, the resulting research into hitherto unexplained and unexplored Scriptures, proved that there were time periods prior to Adam's creation and his expulsion from Eden, and the result of my exploration and research is shown in this Study.

This deals with the time factor, and the Study shows that reconciliation is achievable between Scripture and scientific facts and this resolved part of the dichotomy, which had troubled me.

The next problem to be dealt with was that of evolution, which most people assume to be the only scientific difficulty to be overcome. This is patently not so, but all this caused me to read Charles Darwin's 'Origin of the Species', and H G Well's 'Short History of the World'.

Whilst I found both of these logically understandable, they could not give other than a theoretical basis for the finding of Life forms.

(An interesting anecdote came to my attention recently, which tells of a Lady Hope being at Charles Darwin's bedside, as he lay dying. She tells of him saying "I was a young man with unformed ideas. I threw out queries and suggestions. To my astonishment, people made religions of them". Holding his open Bible, he spoke of the Royal Book and its grandeur, asking Lady Hope to speak of it, commenting, "Is that not the best theme?")

I have studied the evolutionary theories mentioned above in some detail, and it will be seen from the Study that reconciliation between Creation and Evolution is not possible.

There are too many theories of differing, and sometimes opposing points of view to be considered, making any real analysis of evolution impossible. The proven existence of hoaxes, such as the archaeoroptor hoax, (discussed in the Study) makes it extremely difficult to separate theory from fantasy.

As I have said above the theories of Darwin and Wells are logically acceptable, but the existence of Life itself has not been satisfactorily explained. Scientists are still carrying out experiments to try to form life from organic or genetic sources, and so far have not been able to do so.

I feel it is now up to every reader's personal approach to this important subject, as to whether he/she would agree that some light has been shown on the subject, but it is my fervent desire that discussion and criticism will ensue.

Archy T. Gault
October 2003

PREFACE

There is a widely held view that, the deep-seated beliefs of Christian believers in the Creative works of God, and those held by the scientific fraternity of the evolutionary conceptualists, cannot be reconciled. As a result the gap, which does exist between these two ways of thinking, is being allowed to develop and polarize, without consideration being given to the possibility that both views have something to offer.

It is acknowledged that, whilst there are those who hold views which they are not prepared to change, there are some in both fraternities, who do wish for a reconciliation in thinking to be achieved. It is the hope of the author that this study will go some way to assist in this reconciliation being accomplished.

The immutability of the Scriptures is accepted, as is the continuing development of scientific thinking and the author believes that these two principles can be used to bring together the differing beliefs, without causing concern to those who accept that Creation is an undeniable fact. *(1)*

In this study the author has accepted, and has used his belief and understanding, that the veracity of the Scriptures is unassailable. He believes that these are the full revealed words of God and these alone have been the basis on which his research has been carried out. The Scriptures read and studied, are the King James Version (Authorised), and where clarity of translation is sought, reference to the marginal comments from the Scofield and Newberry Editions.

He accepts that the truths in Scripture are complete, and that no further revelation of the mind and will of God will be forthcoming. Jude v 3 reads: **"...the faith** (Gospel/Truth) **which was once (and for all) delivered to the saints"**. This is the basis for his beliefs that no further revelation will be forthcoming. He accepts that the scientific discoveries, which are being announced frequently, are also correct in as far as they go.

It is true to say that the belief in God, the immutability and veracity of Scripture and the Creation as set out above is an act of faith. It can also be said that any moral and ethical questions relating to life, can be answered by fully searching the Scriptures.

This is discussed more fully in Chapter 5.

That the scientific thinking develops from one discovery to another is an obvious statement as, unlike the word of God, new discoveries require assimilation into already formulated thinking and that, as a result, the formulated thinking has to be re-evaluated to accommodate each new discovery.

The belief in the evolutionary theory is dependent upon the assessments of anthropologists on the finds in archaeological 'digs', upon which assessment their thoughts of the form of man are based is also discussed more fully in Chapter 5.

The author has taken a fresh look at the creative acts, as set out in Scripture, and the results of this fresh look, form the subject matter of this study.

The study, whilst based purely on Scripture, takes into account some of these new scientific discoveries as they are tending to show that Creation, as described in the Scriptures, could have happened in a scientific way - not as a result of evolution - but as a result of the Creation in full. That is every creature of every type created by God.

That some of these creatures are now extinct must be accepted and the fossils that have been found show that they did exist at some time in the history of the earth. This is explored at some depth in this study.

The study will show that there were five time periods before the beginning of recorded Biblical time, and these are:

1 The Original Creation. Genesis 1/1 *et al.*
2 Angelic Rule. Ezekiel 28/12-15 *et al.*
3 Void and Dark. Genesis 1/2 *et al.*
4 The Replenishment of the Earth. Genesis Chapters 1,2 and 3.
5 The Creation of Adam. Genesis 1/26-31 and Genesis 2/7 *et al.*

A full chapter will be devoted to the study of each time period. It should be noted here that it is not possible to put any time scale to any of these periods, as Scripture does not give any indication of this, and it would be a presumption on our part to try to do this.

There is one correction, which has to be made to biblical chronology after the above periods are finished, but it is stressed here that this correction **does not** alter the written content of the Scriptures. This correction relates to the acceptance by Biblical scholars of the timescale, first set out by Archbishop James Ussher, in his *Annals of the World* in 1,638 AD, that the Creative work by God was completed in 4,004 BC.

This date is incorrect chronologically, in that, since it is accepted that the Jewish/Israeli recording of their historic existence is almost certainly accurate, and since the Israeli New

Year (Rosh Hashanah) for the year 2,000 AD was 5,761, this makes the start of Israeli history commence at 3,761 BC whilst Ussher dates this as being 1,491 BC. This amendment puts the date of 4,004 BC back to 6,274 BC. This date is also incorrect in the assumption that it was the completion by God of His Creative work, when in historical fact it was the date of the expulsion of Adam from the Garden of Eden.

This date of 6,274 BC is the beginning of recorded Biblical time, and therefore is of great significance.

As stated above, there were five time periods which existed before the recorded Biblical time began and since there can be no date fixed for Adam's Creation and, one date only for his expulsion from Eden, there could have been many thousands, if not millions of years, from the original Creation by God to the expulsion of Adam from the Garden of Eden. (2)

The main difficulty which exists in the reconciling of science to Scripture, and therefore the main point of this study, is that of the extent of time that the earth has been in existence. The study will show that there were possibly millions of years which existed between the original Creation to the earliest historical date of 6274 BC, which can be accepted as being correct and which is the date of Adam's expulsion from Eden.

A critique of the study was given by a friend who was unsure of his belief in God and His Creative works. His comments are valid, and are not of a destructive nature, but show that those of a scientific mindset are willing to discuss this subject. His comments, and the author's responses are given in an Appendix to the Preface and the relevant chapters to which they refer.

A final point should be made here, and that is that although based mainly on Scripture, this study should not be seen as being a Christian study only, as the Scriptures used include the Torah (the Pentateuch) and the Talmud and these form the basis of the Israeli/Judaic religions.

Archy T. Gault

APPENDIX TO PREFACE

(1) Page 1. I regard the fact that the Biblical story is set and inflexible, as a weakness, while scientific theory can change, or have patience to wait for further evidence - this is a strength.

The fact that the Biblical story has remained unchanged, is its strength and not a weakness. It is as complete as it needs to be.

Scientific thinking on the other hand, changes on a regular basis, dependant upon new discoveries, which lead to new theories. Some of these new theories replace those previously held, as being the answer to all of the questions in the scientific mind.

The immutability of Scripture is its strength, not its weakness.

(2) Page 4. The sheer size of the universe, and the fact that we can view objects from which light took millions of years to reach us, proves that it has been in existence for millions of years.

I totally agree with this comment, and the timescale referred to is the main reason for this study being carried out. The Scriptures allow for this type of discussion and thinking to be developed.

CREATION and SCIENCE
RECONCILED

Chapter 1

ORIGINAL CREATION

At the beginning of this chapter let us quote the Scriptures which refer to the occurrence of original Creation and these are Genesis 1/1, where it reads,

"In the beginning God created the heaven and the earth" and John 1/1-3, which reads, **"In the beginning was the Word, and the Word was with God, and the Word was God. The same was in the beginning with God. All things were made by him; and without him was not anything made that was made"**

There are two descriptive expressions in this quotation, which require discussion and these are the use of the word 'beginning' and also the word 'Creation.' With regards to the word beginning, we must ask ourselves, " the beginning of what?"

It cannot mean the beginning of Eternity, as we understand this to have neither a beginning nor an end, but it can be taken from the context in which it was written to mean the beginning of Creation. The word 'Creation' is the noun referring to the verb 'to create', which has the dictionary meaning: 'to bring into existence out of nothing'. (Oxford University Dictionary).

It will assist this study if it can be accepted that prior to the 'beginning', referred to above, there was an existence and that that existence was the God, who is the God of Eternity, and that the two, that is the God and Eternity, are inseparable in our thinking.

It is unthinkable to believe that, before Creation began, nothing existed unless it is taken to mean that nothing of a material nature existed and that the only existence was of a spiritual nature and that this could only be the God of Eternity.

As we progress through this discussion, we will examine the comparison, in thinking terms, between eternity and space and in particular how a created thing, namely time, is used by scientists to measure the extent of space. In scientific thinking the possibility of the existence of the omnipotent being who created all things is excluded. This refers to what they are studying and also to what they use to measure the results of their study.

Let us now consider Time *(1)*.

Time is one of the earliest, if not the first things to be created, and whilst we use it to determine events on Earth, it was also used by the Almighty to describe various events, or occurrences, in the Scriptures. On Earth, time is measured as follows: a day is the time taken by the earth to revolve on its own axis, a month is the time taken by the moon to orbit the earth and a year is the time taken for the earth to orbit the sun. The seasons are also mentioned in Scripture, not as spring, summer, autumn and winter, but as 'seedtime' and 'harvest' and 'cold' and 'heat' and: 'summer and winter, and day and night shall not cease.' (See Genesis 8/22).

Some examples of this usage by the Almighty are as follows:

1 '...woman, mine HOUR is not yet come...' John 7/4

2 '...for the DAY of the Lord is great...' Joel 2/1

3 '...in the midst of the WEEK (of years) ...' Daniel 9/27

4 '...in the beginning of your MONTHS...' Numbers 10/10

5 '...for seasons, days and YEARS...' Genesis 1/14

6 '...until the TIMES of the Gentiles be fulfilled...' Luke 21/24

7 '...which hath been hid for AGES...' Colossians 1/26

From the above, it will be seen that the Almighty used, by divine inspiration through man, the various measurements of time to express His purpose, and for us to study the Scriptures.

That time is used by scientists is patently obvious and it is when he becomes unable to use the normal measures, relative to material things, that he turns to something which is intangible and hidden from all of our natural senses, to measure distances which are beyond human comprehension and that intangible thing is time.

It was Albert Einstein (1879-1955) who first established that time was the fourth dimension, and in his familiar equation of $e=mc^2$, the letter 'c' is used to express time, being the speed of light squared.

That this 4th dimension is necessary is not surprising when we consider the vastness of space and the well known, and much used, measurement of 1 light-year is the perfect example. In lineage terms, a light year is equivalent to 6 million, million miles (6,000,000,000,000), being the distance light travels in one year.

We should look at a simple example of this to explain the need for time to be used as an expression of measuring distance. An example is that our Sun, which is relatively near to us in spatial terms at 93 million miles, emits rays of light which take eight minutes to reach us. It is therefore only eight minutes away from us in time terms. These calculations of a light-year are based on the scientific fact that light travels at the amazing speed of 186.000 miles per second. If the sun is only eight minutes away from us in time terms, consideration should be given to trying to understand the actual distance in miles of the nearest star (other than the Sun) is to us, which is still visible to the naked eye when its distance is 4.35 light-years. At the time of writing, the furthest object seen by telescope is the quasar, known as PK6 2000/330, and this is estimated at being at a distance of 13 thousand, million light-years from us. This is a distance of 78,000,000000,000,000,000,000 miles *(2)*. This is a staggering figure for us even to try to assimilate.

This latter point shows to us that, far from being able to be measured or having a limit, space is unlimited in its extent, and this is far beyond our minds to comprehend. How much more incomprehensible is it that we should discount the existence of the Almighty God who created all this and is there controlling its every movement?

We can now consider that the immeasurable vastness of space gives us an excellent example of Eternity.

Time had a beginning, and it will have an end but Eternity has no beginning, nor will it have an end. Since the purpose of this discussion is to try to determine the Almighty God's plan

Creation and Science Reconciled

for the ages, it would be non-productive to continue this line of thinking and it is best if the foregoing were to be summarized as follows:

- Time is a created thing, and had a beginning, and will have an end.

- Space, to an extent, can be measured and is the continuing source of scientific exploration, and the future alone will reveal the outcome of this exploration. It has no beginning and apparently has no end.

- Eternity is something infinite and, as we have likened it to space to assist us in our comprehension of its extent, it is something which, we must accept as being there. We, as Mankind, having been created in the likeness and image of the Almighty *(3),* were created a living soul (Genesis 2/7) and will share in the experience of eternity with the Almighty. This expression of having: 'been created in the likeness and image of the Almighty' is one, which is open to conjecture, and the simplistic way of explaining this is that man was created as a tri-part being, having a body, a soul and a spirit. The reader's attention is drawn to the comments made by the author in item 3 of the appendix to this chapter, where some conjecture is used to try to answer this more fully.

The Creation of Man is fully discussed in Chapter 5.

Let us now turn our minds to the original Creation as a time period during which all things including fish, birds and beasts were created, or brought into being and we must concentrate our minds on what the Scriptures have to say about the creation. The result of all of the acts of creation was for the pleasure of God and this is borne out by the words of Scripture, **"...for thou hast created all things, and for thy pleasure they were and are created..."** (See Revelation 4/11.)

Returning to the first mention of Creation in the Scriptures, in Genesis 1/1, where it says, **"In the beginning God created the heaven and the earth"**. The heaven referred to here, is not that which is the Almighty's immediate presence and can only be assumed as being something tangible, that is to say, something which can be seen with the naked eye. We get confirmation of this, in Genesis 1/8, where God calls the firmament 'heaven', and to understand the meaning of the word 'firmament' we should turn to the dictionary, where we see that this is: '…the sky regarded as a vault or arch…'. The marginal reference to this in the King James Version (C I Scofield, references) explains it as being: 'the expanse above, the heaven of the clouds'.

The Scriptures also refer to the heavens – plural - as being distinct from the singular heaven and these are metaphorically represented in Scripture as having foundations and pillars (2nd Samuel 22/8), an entrance gate (Genesis 28/17) and windows which opened to pour down rain (Genesis 7/11).

The Scriptures do not give minute details of the heavens, as would normally be required to satisfy the cravings of man's curiosity. Heaven is not merely a state of mind, although it is used as such in normal conversation to describe something which is beautiful, or has heavenly connotations. Heaven is a place, as can be understood by the fact that the Apostle Paul was taken to the third heaven and we read in (2nd Corinthians 12/2) where he, **'…heard unspeakable words, which it is not lawful for man to utter…'** (2nd Corinthians 12/4).

There are numerous references in Scripture which refer to the heavens and it is of import to this study to quote a number of these here.

Psalm 19/1 says, **"...the heavens declare the glory of God; and the firmament showeth forth His Handiwork..."**. The difference made here between 'heavens' and 'firmament' confirms the definition of the firmament being heaven, as distinct from the heavens as set out in Genesis 1/8. The Heavens referred to here obviously include the sun, moon and all the stars/galaxies, all of which were created in the original Creation. (Refer here to the Fourth Day in Chapter 4.)

Psalm 2/4 says: **"...He that sitteth in the heavens..."**

Hebrews 8/1 says: **"...who is set on the right hand of the Majesty in the heavens..."**

There are many more such references, but those quoted above will indicate that Heaven, and the heavens, are a place and places, and whilst we cannot accurately say where they are there can be no dispute that they do exist and that they are not a figment of man's imagination.

It is of supreme importance to point out here that the dwelling place of the Almighty, namely the heavens, are as eternal as God Himself, whilst the heaven (firmament) created in Genesis 1/1 is that part of Creation, which will not exist forever. Revelation 21/1 says, **"...for the first heaven, and the first earth were passed away..."**.

We must now discuss the 'Creation of the earth,' as mentioned in Genesis 1/1, and this part of the study is that which establishes that the earth was in existence for many years, possibly thousand or millions of years, before Adam was 'made' and put in the Garden of Eden.

In order to do this we must look closely at verses 2 to 27 of Genesis Chapter 1, where we are told of the many works of the Almighty, but which in the past have been read but not fully understood by students, scholars and theologians. From verses 3 to 10, God tells us of the acts which restored conditions that had been in existence before. These were acts of preparation for the replenishment on the earth of all created beings, namely birds and beasts, fauna and flora which had existed before the occurrences described for us, in Isaiah 45/18 and Jeremiah 4/23-26.

This replenishment is discussed in Chapter 4.

It should be noted that no mention is made of fish, as they were able to continue to live in the waters, which covered the face of the deep. This state of the earth is discussed in Chapter 3.

Prior to discussing these occurrences, we should look closely at Genesis 1/2, where we will note that, "…**the earth was without form and void; and darkness was upon the face of the deep**…". It should also be noted here, that, although verse 1 includes heaven in the original creation, there is no mention of this heaven in verse 2. There was no cataclysmic occurrence in the heaven of verse 1, this occurrence was only upon the planet earth, and it would be both Scripturally and spiritually, wrong to assume that God would create something which was 'without form and void' for his pleasure. Refer here to the Scripture quoted in this Chapter, namely from Revelation 4/11.

It is relevant to the above, and to the development of this study, that Isaiah 45/18 be quoted in full here, and it says, "**…God himself that made the earth and formed it: he hath established it not in vain, he formed it to be inhabited: I am the Lord; and there is no one else…**"

This Scripture alone, is sufficient to show that the earth was not created as being 'without form and void', and it will be significant for us to read, study and consider, what is written in Jeremiah 4/23-26. There it is written, **"...I beheld the earth, and lo, it was without form and void..."** which are the identical words of Genesis 1/2 and, since it is not known if the words of Moses who wrote the book of Genesis, had been available to Jeremiah, it would be fair to assume that they were written by Jeremiah under divine inspiration.

Jeremiah goes on to say, **"...I beheld the mountains, and lo, they trembled, and all the hills moved lightly. I beheld, and lo, there was no man, and all the birds of the heaven were fled.... all the cities, thereof, were broken down at the presence of the Lord..."**.

These Scriptures indicate to us that the earth had undergone some serious disruption between verses 1 and 2 of Genesis 1 and again, Isaiah 24/1 shows, that whatever happened, had been at the hand of the Lord. This verse says, **"...Behold, the Lord maketh the earth empty, and maketh it waste, and turned it upside down..."**.

We may try to understand what the earth was like in its original form at the original Creation of and even in its present state we find it to be a world of such beauty. *(4)* With mountains, valleys, rivers, seas, and coastlines which, together with the fauna and flora with which it abounds, it is beautiful, and it would be beyond our ability, to see in our minds eye, the form that it had when it was perfect. Nothing less than perfection would satisfy the Lord God, who had created it for His pleasure.

That it had in its perfection, cities and inhabitants, is shown in Jeremiah 4/24 and 25, but even at the time of its upheaval, the Lord God states, **"...yet I will not make a full end...."** (Jeremiah 4/27), showing that, even when cast into such an empty state, the Lord God had plans for its future use.

That the world had had inhabitants, and other created fish, birds and beasts, is clearly shown in the verses quoted above. The statement that the birds had gone shows that there were creatures which lived on earth, and that cities had been there, but had been broken down but that there was no man. This comment regarding there being no man, will be dealt with later, in Chapter 2, but for the moment let us remain with the comment about the birds of the air.

It is true to say that after the catastrophe which occurred, all of the birds of the air, and all land animals which had existed prior to the cataclysm, ceased to exist.

As mentioned, the sea creatures were not destroyed, and evidence of this is seen in the fossil of the coelacanth, which we will study on later.

Having established from Scripture that the original Creation of the earth took place, and that the earth, heaven (as distinct from the heavens) and time are created things, let us now examine what was in existence before these creative acts of the Almighty took place.

This can only be the existence of the Almighty God Himself. Again, we must turn to Scripture for evidence of that, and this evidence can be found in various parts of Scripture. We will include quotations from two Scriptures to show the existence of the ever present Being.

The first of these references is in Exodus 3/14, where Moses was commanded to say, when asked: "…who sent you…?", his reply was to be that: **"…I AM hath sent me…"**, this quotation being prefaced by the statement by God to Moses, **"…I AM THAT I AM…"**. This statement by the Almighty signified the lack of any possibility of there having been any change in the existence of the Almighty God.

Creation and Science Reconciled

(At this point, before looking at the second reference, it will be useful to know why Moses was in the position he found himself of being asked by the Almighty to undertake the mission which God had planned for him.

The Children of Israel were being held in captivity in Egypt, and were being seriously oppressed. Their cries for help had been heard by God and Moses, having been born to a Hebrew mother and father but raised as the 'Son of Pharaoh's daughter' had, at the age of 40, fled from Egypt. He had spent 40 years in the Land of Midian, and had stood aside to see the strange sight of a burning bush, which was not consumed by the fire. It was from here that the Lord God spoke to Moses, and told him of his mission to free the Children of Israel from their bondage.)

The second reference is made by Jesus, who, when he asked of the men who had come to take him, "…Whom seek ye…?", and they told him who it was that they sought, his answer was **"…I AM HE…".** Upon hearing these words, they all fell backwards. See John 18/4,5 and 6.

It is important to note in the falling backwards (verse 6) of those who heard the voice - uttering the title of the ever present being – that, as in normal worship, it is usual to fall forwards to be in obeisance to the person being worshipped. Their falling backwards signifies their unawareness of the Being to whom they were speaking.

To conclude this chapter in our discussion, let us accept that we will never, whilst in our bodily state, be able to understand eternity but that we must accept, from Scripture, that there is an ever present Being who created all things, and who has been in existence since before time began.

It will be obvious to all readers, that it is impossible to try to put any time scale on the events discussed above, and suffice it to say that ***these events did take place***.

As a footnote to this chapter, we should note that any reference to the Lord God has been confined to the ALMIGHTY, the LORD GOD, GOD and I AM, since it would be premature to introduce other titles.

APPENDIX TO CHAPTER 1

(1) Page 10. I find your distinction between eternity and time interesting. The only thing I can conceive of being eternal is 'time'. I can't accept that a thing, or being existed eternally and had no beginning - my brain will not compute that.

There is a paradox for you to consider - if God has existed eternally, he can't possibly have waited eternally to begin Creation. There must have been other Creations - infinite numbers of them; eternity allows for an infinite number of Creation processes. But it is not possible for an infinite number of anything to exist.

There is a lot in this comment.

When time is considered as being part of the Biblical story for earth, it cannot be regarded as being eternal. However, if time is considered as part of the universe as a whole, then there is every possibility that in other constellations, time does have an eternal element to it. Here on earth it is governed, and dictated to, by the planetary rotations of the earth, moon and sun around one another.

I agree that it is difficult to accept that there was a Being who existed eternally, but there is no real paradox here, as we do not know how long God waited, before He began the Creative works.

The use of the phrase 'waited' implies time, but in this study, which is purely earth related, eternity has to be accepted as being part of the Biblical narrative.

It is true to say that there may have been other parts of the Creation, but not other 'Creations', as all other 'Creations' would have formed part of the 'Original Creation,' and there may indeed be an infinite number of these Creations.

This may be an important point, scientifically, but this study looks at the effect of the Creative acts from the point of view of earth.

To say that it is impossible for an infinite number of anything to exist is not mathematically correct, since the square root of the number two is, in itself, an infinite number, and this forms the basis of the Pythagoras theorem. This theorem is probably the most significant discovery in the realm of mathematics, as it forms the basis upon which many other such theorems are built.

(2) Page 12. The light from the quasar took 13 thousand million years to reach us, so Creation couldn't happen later than that, so you know that there are definitely millions of years between Creation and Adam.

I totally agree with this comment.

I believe that space is unlimited, because it is impossible for nothingness to exist.

I totally agree with the first point here, but would add that the 'nothingness' is an apparent 'nothingness', and is moreover the existence of the Eternal God.

I am adding a quotation here, from a statement written by an acquaintace, which reads as follows:

"When life has lost all meaning, and we reach the nadir of our existence, we can adopt the attitude that there is nothing of any import left in life.

The complete absence of feeling, any feeling, has got to be experienced before it can be fully understood; being cold or warm - it does not matter; hungry or not; loved or unloved; alone or in company, these are all of no importance and the very act of breathing is of no significance;

nothingness is all that is important. The nothingness was of such import as it became everything that I had...the nothingness that I felt was, in fact, everything that I needed and that all else would be given to me through...the Holy Spirit of God"

Relative to the statement that 'Time had a beginning...' - a debatable point. I think time would go on, even if this universe disappeared.

I cannot agree with this point as, if the universe ceased to exist, why would there be a need for time? It is a very important point to consider and it may be an assumption on my part that time was a created thing but it remains true to say that, as far as earth is concerned, time began when the creative works began, and that it will cease when God's plan for the ages has been accomplished.

(3) Page 13. Relative to 'in the likeness and image of the Almighty,' - I have never understood this phrase. Does it refer to a physical likeness? - That would mean that God had legs and a mouth - why would he need a mouth? - Does he eat?

It must be remembered that God is a Spirit, and as such does not have parts, which could be easily identified as being of a physical nature. My understanding of these expressions is again based on Scripture, in that being created in the 'likeness of God' would be shown by man's tri-part makeup of body, soul and spirit.

In the 'image of God' is much more difficult to understand, but an interpretation could be that as man was created perfect and given dominion over the earth and all created beings, he was in the image of God who is perfect and has dominion over all of Creation, the earth and all of the universe.

Archy T. Gault

(4) Page 17. Was the planet earth created at the start of 'Creation'?

I think that this planet is a lot younger than the quasar. I have heard a figure of 6,000 million years for the age of the earth.

My thoughts on this are that, as we are not given the sequence of the creative acts, we cannot be sure if the earth pre-dates or post-dates any other act of Creation. The time scale of 6,000 million years as the age of the earth is one of the points that scientific thinkers, and those who are prepared to read this study, could perhaps agree as being possible.

Chapter 2

ANGELIC RULE

It can be seen from the previous chapter that, prior to the cataclysmic occurrences which took place on earth, there had been a habitation on earth and that this had included birds, beasts and fish.

Let us now consider, and discuss, what this habitation had consisted of in addition to those creatures mentioned above. Let us again turn to the Scriptures for the information necessary to confirm what this was and we will find part of the answer to this in Ezekiel 28/12 to 15.

That it was of angelic form can be shown, and that it was under the governance of Satan, can also be shown, all of course under the direct control and rule of the omnipotent Lord God Almighty.

Before discussing this aspect of the angelic rule, it will be found useful to study the being of Satan, and this is clearly expressed in the Scriptures mentioned above. It is important to quote these in full and they are as follows (the underlined words being the subject of our further discussions).

"...Thou sealest up the sum, full of <u>wisdom</u>, and <u>perfect in beauty</u>. Thou hast <u>been in Eden</u>, the garden of God; every precious stone was thy covering, the sardius, topaz and the diamond, the beryl, the onyx and the jasper, the sapphire, the emerald and the carbuncle, and

gold; the workmanship of thy <u>tabrets</u> and of thy pipes was prepared in thee in the day thou <u>wast</u> <u>created</u>. Thou art the <u>anointed</u> <u>cherub</u> that covereth; and I have set thee so; thou wast upon the <u>holy</u> <u>mountain</u> <u>of</u> God; thou hast walked up and down in the midst of the stones of fire. Thou <u>wast</u> <u>perfect</u> in thy ways from the day thou wast created, till iniquity was found in thee...".

This is the picture of Satan as a created being and we can assume, from Scripture, that there were many angels under his authority. Jude v 6 refers to these, as follows,

"And the angels which kept not their first estate, but left their own habitation..."

It can be accepted that their first estate was that which was given to them at their creation, and this can also be accepted as being conterminous with the creation of Satan.

The habitation referred to here was on earth. The reference to the Garden of Eden and the holy mountain of God is earthly, but no doubt it can have heavenly and spiritual connotations.

Let us examine the 'personality' of Satan from the information given to us by Ezekiel.

1. He was a created being and was full of wisdom (but not omniscient) and perfect in beauty.

2. He was in a position, or estate, of being in authority (but not omnipotent) by being the anointed cherub that covereth.

3. He was perfect in his ways until iniquity was found in him. When iniquity was found in him he was cast out, but no reference is found that he lost his wisdom or beauty, neither did he lose his authority over the angels who fell with him.

Creation and Science Reconciled

The finding of this iniquity in him was the reason for the serious disruption brought to bear on the earth. That this was brought about by the Almighty God, is discussed in some detail in Chapter 1 and the Scriptures, which give the details of Satan's iniquity, are quoted here from Isaiah 14/12-14:

"How art thou fallen from heaven, O Lucifer, son of the morning! How art thou cut down to the ground, which didst weaken the nations! For thou hast said in thine heart, I will ascend into heaven, I will exalt my throne above the stars of God: I will sit also upon the mount of the congregation, in the sides of the north: I will ascend above the heights of the clouds; I will be like the most High."

As he did not lose his wisdom, he is to be feared by believers today, because he has a deeper understanding of the Scriptures than we have. We must give thanks that although he is full of wisdom, he has not the power of being able to foresee the timing of the prophetic writings anymore than we have.

His creation, as touched on above, was as an eternal being, as is said by the Lord and referred to in Luke 20/36, where it says:

"…Neither shall they die any more, for they are equal unto the angels…" of which Satan was the 'anointed cherub that covereth,' and over which he had dominion.

It will be useful to consider at this point that although Satan is given the male gender, this does not mean that there were two genders in the angelic creation and, in fact, Scripture states that angels were not given in marriage, (see Luke 20/35). They were not given the ability to 'be fruitful and multiply'.

The popular and sentimental view, which many people have of an angel being of a lovely female form, with wings, is not borne out by Scripture.

The Scriptural statement that angels: **"...neither marry, nor are given in marriage..."** Luke 20/35 negates the view held by some believers that the 'sons of God' referred to in Genesis 6/2 were angels.

When we consider his position of authority it will be confirmed that his dominion was over things relative to the earth. This can be seen by the reference to **'Eden'** in verse 13, to the **'holy mountain of God'** in verse 14, both being physical places, and on earth. That this dominion extended to the birds, beasts and fish can also be accepted, with particular emphasis being given to Jeremiah 4/24 and 25.

Let us for a moment explore the pre-Adamic existence of each species. Fossil remains of bird life have been found on earth, and these are of pterodactyls, which have been carbon dated by scientists as having been on earth some 70 million years earlier. Fossil remains also exist of beasts, such as dinosaurs, for example, brontosaurs, and these have also been carbon dated as being on the earth some 70 million years earlier.

Fish fossils have also been found, particularly that of the coelacanth, which was thought to be extinct until 1938, when some 100 living specimens were caught off the coast of Madagascar. The carbon dating of the fossils also dates them at about 70 million years old.

It could be assumed that similar dates given to these is coincidental, but the important fact is that there were birds, beasts and fish which were in existence on Earth during that period of time. To assume otherwise would be to fly in the face of scientific knowledge, and that is not the purpose of this study, but rather to show that Scripture confirms the existence of a habitation on earth; possibly many millions of years before man was created. It would be a safe conclusion to reach that this habitation was in existence on the earth at the time when Satan had dominion over the earth.

Let us now continue our study by looking at the downfall of Satan, and again we must turn to the Scripture for the accurate account of this and the reasons for it. In Ezekiel 28/15 it is written that,

"...Thou wast perfect in thy ways from the day thou wast created, till iniquity was found in thee..." Verse 16 of this chapter goes on to say **"...therefore will I cast thee as profane out of the mountain of God, and I will destroy thee, O covering cherub....Thine heart was lifted up because of thine beauty, thou hast corrupted thy wisdom by reason of thy brightness...thou hast defiled thy sanctuary by the multitude of thy iniquities...".** Isaiah 14/12-15 gives further information on his downfall, it says: **"...How thou art fallen from heaven, O Lucifer, son of the morning...for thou hast said in thine heart...I will sit upon the mount of the congregation...I will be like the most high...".**

These Scriptures show that the casting out of Satan had a twofold aspect in that he was fallen from heaven and was cast out from the earth and that the place of habitation for him and his host of fallen angels was in the air, which surrounded the earth.

Let us examine the names by which Scripture refer to him, and some of these are:

Prince of the power of the air...Ephesians 2/2
Prince of this world...John 12/31
Prince of darkness...Ephesians 6/2
God of this world...2nd Corinthians 4/4
Serpent...Revelation 20/2

This latter name reveals that one of the powers which Satan had, and possibly still has, is to enter into the body of another being. This is borne out by an examination of Genesis, Chapter 3/14, where it will be seen that, prior to the temptation of the woman by Satan in the form of a serpent, the serpent was a beautiful creature and walked upright. If this were not so God would not have said to him, **"…upon thy belly thou shalt go…"** This confirms that Satan had the power to take on the physical form of another being.

All of the above led to the cataclysmic occurrences described by Jeremiah and Isaiah, and which we discussed in Chapter 1, and we shall see the outcome of the cataclysm in Chapter 3, 'Void and Dark', but we must remember the words of the Lord God, spoken through Jeremiah in chapter 4/27, **"… yet I will not make a full end…"** as this gives us the hope and promise, that the Almighty had plans for the earth and its future inhabitants.

Chapter 3

VOID AND DARK

The title to this chapter is a paraphrase of the description of the earth, after the cataclysmic occurrences discussed in the previous Chapter *(1)*

It is important that we quote here, in full, the Scripture which has been paraphrased in the title to this chapter, and this is to be found in Genesis 1/2, which reads:

"And the earth was without <u>form</u>, and <u>void</u>; and <u>darkness</u> was upon the face of the deep. And the Spirit of God moved upon the face of the waters".

This quotation is from the Kings James Version, and it will be useful to compare the above with the quotation from the Revised Version of 1884, which reads:

"And the earth was waste and <u>void</u>, and <u>darkness</u> was upon the face of the deep".

This comparison highlights the difference between 'without form' and 'waste and void' and it would be true to say that the Revised Version quotation is the more accurate, since the earth would still have retained the form in which it was originally created. Both quotations refer to the 'darkness', which was 'upon the face of the deep.' That it was waste is without doubt, and that darkness covered the face of the deep, is also without doubt.

'The face of the deep' here shows that the entire surface of the earth was covered with water and, as a result, all the flora and fauna became extinct. Further discussion in chapter 4 will show that fish life still remained in the waters. Reference is made to this in chapter 2, where note is taken of the existence of the fish form coelacanth.

The quotations above are repeated in Jeremiah 4/23, and are fully discussed, in detail, in Chapter 2.

Confirmation that the rule of the Almighty over the earth remained whilst it was 'waste and void', is seen in the latter part of Genesis 1/2 where it states that 'the Spirit of God moved upon the face of the waters', showing that His control was still evident during this time period. This continuation of the omnipotence and control over the earth, is commented on by the Psalmist in Psalm 104/30, which reads:

"Thou sendest forth thy spirit, they are created: <u>and thou renewest the face of the earth</u>".

This quotation has an historic reference as is seen in the first highlighted part, and also a prophetic content, in that the renewal of the earth is, and will be, an ongoing process. The 'renewal' reference can also be taken to refer to the replenishment of the earth, which will be discussed in Chapter 4.

As a precursor to the replenishment of the earth, it is significant to note that at no time in Scripture does it say that the heaven of Genesis 1/1 was waste and void, and we can safely assume that this heaven was not affected by the catastrophe, which occurred on earth.

Since the heaven was not affected by the downfall of Satan, we can be assured that the sun *(2)* still existed, having been created in the original Creation and its light was prevented from shining on the face of the earth by the cloud cover, which did occur as a result of the catastrophe. This cloud cover stopped the life supporting rays from reaching the earth, and prevented the flora and fauna from surviving. The seed from these two forms of material life did not die, but lay dormant under the waters, which covered the earth. This point will be developed further in chapter 4, when we discuss the replenishment of the earth in preparation for the Creation of man.

This part of our study is of necessity brief, as we are told no more than is quoted above in Scripture. To try to develop this as conjecture, or speculative, would be counter-productive and go against the purpose of our study. Finally, it can be said that it is not possible to put **any timescale** to this period.

APPENDIX TO CHAPTER 3

(1) Page 31, para 1. My interpretation of these events would be a huge asteroid hitting the earth.

My answer to this point is that, as Scripture is silent on it, any assumption could be valid but, since the study is Scripturally based, it would be wrong to include any such speculative thinking.

(2) Page 32. Scientists say that the sun is a young star, compared to many others. Stars are forming and dying all the time.

Scientists say many things today and different things tomorrow. I refer to my response on the immutability of Scripture and the changing face of science, under Item 1 of the Preface.

Can you be sure our Sun was in at the Original Creation?

The simple answer to this question is a straightforward **YES**.

Chapter 4

THE REPLENISHMENT OF THE EARTH

Before commencing our study in detail into the 'Replenishment of the Earth', we should confirm that the Lord God Almighty *did* carry out this work in six days. This we can accept without any compunction although there is continuing talk in the world at large, as to whether the 'day' mentioned in Genesis relates to a day of twenty-four hours, which seems to be generally accepted as impossible. We should refer to the Scripture in 2nd Peter 3/8, which reads: **"...that one day is with the Lord as a thousand years, and a thousand years as one day..."** It is, therefore, irrelevant as to what the extent of time to which the 'day' in Genesis refers, but what is relevant is that, to the Lord God nothing is impossible, as is stated in Scripture in Luke 1/37, which reads: **"...For with God nothing shall be impossible..."**

What is important is that we accept that the replenishment of the earth *was* carried out by the Lord God Almighty, and the argument as to the time extent of the 'day' is of no importance.

We have already discussed the original Creation, in Chapter 1 and have looked at the events which occurred prior to the need for the earth to be replenished. We should now proceed with our study into the manner in which this replenishment was done, and our task will be made easier if we look at and discuss these works on a day-to-day basis.

THE FIRST DAY

It is helpful to quote here in full, the Scripture which tells us what happened, and this is to be found in Genesis 1/3-5, which reads: **"And God said, let there be light, and there was light. And God saw the light, that it was good: and God divided the light from the darkness. And God called the light Day, and the darkness he called Night. And the evening and the morning were the first day".**

It is here that we must ask ourselves the question "where did the light referred to come from?" There can only be one answer to this question, and that is that it came from the sun, which had been created in the original Creation.

The commandment for this light to appear was for God to order the clouds, which covered the face of the earth during the period of 'Dark and Void' (which we discussed in Chapter 3), to dissipate to allow the light from the sun to penetrate through to the surrounding atmosphere. This does not mean that the sun was then visible from the earth, as this did not appear until the 4th day. We can understand this, even as we sit and look out at the countryside on a cloudy day as we are seeing the light without seeing the source of that light.

We must also think back to the way the earth was during the period of 'dark and void' when **"...darkness was upon the face of the deep..."** This darkness must have been caused by a deep and impenetrable thickness of cloud, which prevented the very light of the sun from breaking through.

This divine instruction was for the clouds to allow the light to penetrate and filter the atmosphere for the purpose of allowing the natural and created act of the earth's rotation on its own axis to divide the light from the darkness, as instructed by God in Genesis 1/4. The calling of the light 'Day', and the darkness 'Night' is a further evidential fact of the divine control over every aspect of the universe.

The commanding of the light to appear is, in itself, supportive of the fact that the universe had been created in the distant past, as we discussed in Chapter 1. We will study the depths of meaning in the opening up of the cloud layers, to enable the sun and moon to be seen from the earth, when we discuss the Fourth Day later in this chapter.

Let us look at a significant statement before moving on to look at the Second Day and this is that the apparent start of the present usage of the word 'day' was in the evening which is contrary to what we would expect as we consider the start of each day to be the morning. Scripture is in fact perfectly correct in this difference in that the actual start of the measurement of each day, as we know it, is at midnight or as the Scripture says 'evening'. It is also significant, that in this day and age the Hebrew day does still start at sundown, as can be exemplified by the truth that the Hebrew Sabbath starts at sundown on Friday and lasts until sundown on Saturday.

THE SECOND DAY

Let us quote here from the King James Version, where it says, **"...Let there be a firmament in the midst of the waters, and let it divide the waters from the waters..."** *(1)*

The marginal reference for this passage states that the meaning of this is an expanse of air between the waters beneath and the vapor above.

The 1884 Revised Version repeats the quotation noted above and, in a footnote, states that the firmament referred to is the Hebrew word for an expanse. The Thomas Newberry Edition of the King James Version has the same quotation and also has the marginal reference of the firmament as an expanse. The New International Version of 1978 says, and here let us quote it direct as the modern translation can give us a more readily acceptance of the meaning: **"...Let there be an expanse between the waters to separate the water from the water...and God called the expanse 'sky'..."**

In order that we can discuss this in a meaningful manner, we must take each of the renderings above and look at them objectively. In doing this, we can see that there is no real difference between them and that they all agree that an expanse was formed to separate the water above from the water beneath and that the water above is referred to as vapor.

Before proceeding further, let us accept that the cloud which formed the darkness; **"which was upon the face of the deep"** had a cloud base of zero, in relation to the earth's surface. At that time this surface was completely of water and the density of the cloud was of such intensity that, prior to the commandment: **"Let there be light"**, it did not allow the light of the

Creation and Science Reconciled

sun to reach the surface of the earth.

The expanse which was formed created a space between the waters and this space **"God called sky"**. (New International Version quotation) This created a higher cloud base than had existed before.

The Scriptures quoted above substantiate that there were clouds of water in vapor form, and these Scriptures also substantiate that they were separated from the waters which were below. That these events were carried out is confirmed by the words of Genesis 1/7, where it says **"...and it was so..."**

The Second Day finishes with the words **"And evening and morning were the second day"**.

THE THIRD DAY

We turn here to Genesis 1/9-12, and can see from these verses that there were two acts of replenishment following on from those of the Second Day and each is supportive of the fact that it was a replenishment which was taking place and not acts of original Creation.

The first act which we should consider is set out in Genesis 1/9 and 10, where we read that the command from God was for the: **"waters to be gathered together... and for the dry land to appear..."** *It does not say here that the dry land was created but was made to appear, confirming that it had been in existence before being covered by water. The waters being gathered together into one place to permit the land to appear would also confirm our previous thoughts that the land had been there, but had been* *covered by the waters during the cataclysmic occurrence which took place at the downfall of Satan.*

The second act which is of great significance, is that which is expressed in Genesis 1/11, and is of such import that we should quote the actual words of the Scripture: **"…Let the earth bring forth grass, the herb yielding seed, and the fruit tree yielding fruit, after his kind, whose seed is in itself, upon the earth…"**

Let us look at two important statements here, firstly, in the use of the words, **'bring forth,'** and secondly, **'whose seed is in itself, upon the earth.'**

The words **'bring forth'** do not infer an act of Creation, but rather an act of producing something which is already there. The second statement provides us with the answer as to what was to be brought forth, and that was from the, **'seed which is <u>within</u> itself, upon the earth'** which was the life form of the grass, the herbs and the fruit trees. That this life form had lain dormant in the earth for a vast period of time is acceptable to us, as we can see this type of event occurring after each flood, or period of drought in the deserts of the earth. The seeds can lie dormant being brought forth into bloom once light, dry land or water are made available.

We have seen that in the First Day God commanded the light to appear, in the Second Day He made the expanse of air to be formed by raising the cloud base to a level which would allow the air to form, and in the Third Day He caused the dry land to appear.

Once again, the account of Third Day is ended by the words **"…the evening and the morning were the third day…"** but here for the first time in our study of Genesis 1, we read the words **"…and God saw that it was good…"**

This statement is made after each of these two acts of replenishment in the Third Day are completed and the significance of this cannot be overlooked. We must ask ourselves why this statement should be made here, and not in the two previous days. The answer must surely lie in the fact that out of these two acts, and supported by the acts of the First and Second Days, God once again saw that Life was upon the face of the earth.

THE FOURTH DAY

In considering the Fourth Day we can see a difference in the terminology which we must consider in order to ask why this occurs, as it does affect our understanding of the results of their use.

The main difference occurs in Genesis 1/14 and 15, where we read that God said: **"...Let there be lights..."** and **"let them be for lights..."** as these expressions do not infer, or imply, a creative act. They are commands for what was already in existence that the **'lights'** were to be able to be seen from the planet earth. Verse 14 commands the lights to separate the day from the night, noting as we do, that it was not to separate the darkness from the light, as is stated in verse 4, but is in reverse order to divide the day from the night. The command was also for the lights to be for signs, and to enable the days, seasons and years to be measured. For this we should refer back to our discussion of the original Creation in Chapter 1.

In verse 15 we read that the command was for the lights to give light upon the earth and this light should not be misconceived as being the same light as was discussed in our study of the First Day, as this light was directed to **'shine upon the earth.'** The light referred to in verse 14 refers to the **direct** sunlight and moonlight, which can be seen today, and which have more life giving powers than the indirect light, which was discussed in our study of the First Day.

It is now, that we must consider the true meaning of the differential terminology. When in verse 14 it says, **"... and God made the two great lights..."** referring to the sun and the moon, the word **'made'** does not imply a creative act, but simply a command for the existing light to be able to be seen from the earth. This view is supported by the words at the end of verse 16, which are, **"...and he made the stars also..."** It is understood that the word 'made' used here for the sun, moon and the stars, has a different meaning from the word 'create', in the original language of the Scriptures, and here it means 'made visible' or 'made to appear'.

We have previously discussed the fact that the planet earth had been affected by a cataclysmic occurrence, but here the reference to the stars shows of their unaffected existence from the original Creation.

In verses 17,18 and 19 we find confirmation of the carrying out of the commands of the Almighty, given in verses 14 and 15 and again we read the words that God was pleased with the works. As in verse 18 we read, **"...God saw that it was good..."**

Again it says: **"...And the evening and the morning were the fourth day..."**

That the earth was now ready to receive and support further life forms is evident by the acts referred to above. Let us now summarise them, before proceeding to study the Fifth Day:

Creation and Science Reconciled

Act 1 was to permit light to penetrate the hitherto impenetrable layer of cloud which covered the earth.

Act 2 was to form an expanse between the waters below and the expanse above.

Act 3 was to order the dry land to appear and to be separate from the waters, or seas.

Act 4 was the bringing forth of the grass, herbs and fruit trees.

Act 5 was to order the direct rays of the sun to reach the earth's surface, thus enabling the earth to support the higher life forms which God had planned for it and which we will discuss during our study of the Fifth Day.

THE FIFTH DAY

We will see here again the use of the words **'bring forth',** the words which were looked at in some detail in the Third Day and in this instance they are used in the **'bringing forth,'** by the waters of the life which they contained.

Let us quote here the relevant Scripture, from Genesis 1/20, which reads: **"And God said, let the waters bring forth abundantly the moving creature that hath life."**

A close examination of this passage will show that the bringing forth was to be abundant and also that the creatures to be brought forth had been in the original existence. This quotation does not say that the moving creature will have life; it says simply the creature which hath life. There can be no other meaning to be taken other than the life which was commanded to be brought forth was already there. This deals with the creature content of the seas; creatures not made extinct as a result of the cata-

clysm which had occurred. (Refer here to the existence of the coelacanth, discussed in Chapter 2.) This is apart from the great whales which were created here (see Genesis 1/21 for this) and this new Creation would have been due to the fact that the great whale, and similar mammal species, were and are, dependant on air for their life to be maintained and this air was not formed until the Second Day of the replenishment of the earth.

This passage of Scripture also refers, in verse 21, to, **'every living creature that moveth'**, and **'every winged fowl'**. All of these were to be **'after his kind'**. This statement, **'after its kind'**, is of such importance today when considering the fertilization and cross breeding of different breeds of animals that it must cause us, as believers, to be made fully aware of the disobedience of man to the divine will and command of God.

The statement **'after his kind'** precludes the possibility of any evolution from ever having taken place, since every example of **'each kind'** was formed in the same day of replenishment, and each kind includes the primates *(2)* from which, evolutionists, theoretically, say man evolved.

The scientific world continue to search for the 'missing link' to try to prove what is, in effect, only a theory and to the believer this 'missing link' will never be found, as it never existed.

God blessed the creatures, by saying, **"Be fruitful and multiply, and fill the waters in the seas, and let fowl multiply in the earth."**

Again we read that, God saw that it was good (verse 21) and that evening and morning were the Fifth Day. (verse 23)

THE SIXTH DAY

Let us again quote from the King James Version for this, which was surely the ultimate purpose in the Almighty's plan for the replenishment of the earth, namely the Creation of man.

The quotation reads, **"And God said, Let us make man in our image, and after our likeness; and let them have dominion over the fish of the sea, and over the fowl of the air, and over the cattle, and over all the earth, and over every creeping thing that creepeth upon the earth."**

When we study what is said here, we can see it as being the very hub of our existence, in that, the Creation of Man was the work of the Triune God and, being in the image of God, man was created with a three-part make up, having a body, a soul and a spirit.

When we look at the soul of man, we should consider the words in Genesis 2/7, where we read that God, **"...breathed into his nostrils of the breath of life; and man became a living soul"**.

The meaning of the word 'living' is that of an undying soul. This soul is the essence of the being of the person and the soul will live forever, in either a state of being saved or unsaved, either in Heaven or in the 'Lake of Fire'.

When a man dies his body is put into the grave where it will fall into corruption (see 1st Corinthians 15/42) and **"...for dust thou art, and unto dust thou shalt return,"** as stated in Genesis 3/19.

When death occurs the third part of his being returns to God (see Ecclesiastes 3/21) where we read that the spirit of man **'goeth upward'**.

Man was given dominion over all of the replenished earth and over every creature and it should be noted here that there would have been no fear of man in the animal kingdom as everything was replenished in a state of perfection.

We are given further information as to God's purpose in the Creation of man, when we read, in Genesis 1/27, that, **"...male and female...created He them"**. The purpose of the Almighty in doing this is shown to us in Genesis 1/28, where we read, (and it is of such importance to our study that once again we should quote directly from Scripture) **"And God blessed them, and said unto them, be fruitful, and multiply, and replenish the earth, and subdue it..."** This is the first recorded commandment of the Almighty to Man and shows that, since God had replenished the earth with everything which had been in existence prior to the downfall of Satan, the replenishment here must surely mean the filling of the earth with beings which had not existed before, and these beings were to be the offspring of the man and the woman.

There can be no doubt that the man and the woman fulfilled this commandment, as not to have done so would have brought about the Fall of Adam at a much earlier time than that in which it actually occurred, namely at Adam's disobedience regarding the commandment not to eat of the **'fruit of the tree of knowledge of good and evil'**.

Satan had never been given this ability to procreate and in all probability this could be the main cause of his hatred of Adam, together with man having been given Satan's original position of having dominion over the earth.

Let us return to the creation of Adam, as this word 'Creation' has a different meaning to that which we referred to in Chapter 1, when we saw that the meaning there was to bring into being, out of nothing, something new. In this instance we should look to Genesis 2/7, where we read: **"...the Lord God formed man of the dust of the ground..."** This creative and formative act by God brought into existence a being which was in the image of God and, in His likeness, had privileges and responsibilities which had not ever been given to any other created being. He was made from the dust of the earth, and when dead will return to this state.

It is worthy of note here that Adam and the woman were to eat of the herb and the fruit of the earth and it would be true to say that, in his original state, man was vegetarian in his diet.

The ending of the Sixth Day is recorded as before, with the words that, evening and morning were the Sixth Day, but there is a wonderful expression which did not occur before and that is when God saw every thing that he had made: **"<u>behold it was very good</u>".** (Genesis 1/31)

By referring to the opening paragraph of this chapter, it will be seen that, *it is not possible to put any timescale* to the events discussed, in each of the 'days' of this 'Replenishment of the Earth.'

APPENDIX TO CHAPTER 4

(1) Page 38, Second Day. This could be interpreted as meaning a continent was formed separating the oceans?

This cannot be accepted, as the reading of Scripture clearly states the separating of the water beneath from the vapor above. This is covered in greater detail in Page 3 of this chapter.

(2) Page 44, Fifth Day. I don't think the theory says humans evolved from primates; rather they could have had a common ancestor.

My understanding of evolution is that man is said to have evolved from a primate, but whether man and a primate descended from the same ancestor, is contrary to Scripture. Man was created as man, with no predecessor.

There are many missing links, some of which have been found recently.

If these 'missing links' have been found, and are seen as such, why are they still called 'missing'? Why is the search continuing for other 'missing links'?

Chapter 5

THE CREATION OF ADAM

Immediately following the replenishment of the earth, discussed in Chapter 4, the earth was now ready for God's final act of creation, and that was the creation of Man, who was created in the image and likeness of God. We discussed this, in part, during the 'Second Day' in Chapter 4, and now can develop this in some detail.
- Man, thus created, was perfect in every way having been created in the image of God.
- Woman was perfect, having been created from within the perfect man.
- The fish of the seas, the fowl of the air and the beasts of the earth were all perfect, having been created by Almighty God in His replenishment of the earth.

The Edenic condition is referred to in modern language as being Paradise and there is Scriptural support for this reference. This is to be found in Revelation 2/7, where we read that,

"**…the tree of life, which is in the midst of the paradise of God**", and we also have reference to the tree of life in Genesis 2/9, where it says "**…the tree of life also (stood) in the midst of the garden…**" so it can be safely assumed that the Garden of Eden can be called, and was, Paradise.

There is more to the subject of Paradise than is contained in the Edenic reference, and this is a subject which will be discussed in another study by the author.

Let us return to the state of perfection, which we touched on above, as this is set out for us in Isaiah 11/6 and 7 and, because of its importance, it should be quoted here in full. It reads,

"...the wolf shall dwell with the lamb, and the leopard shall lie down with the kid; and the calf and the young lion and the fatling together, and a little child shall lead them. And the cow and the bear shall feed; their young ones shall lie down together; and the lion shall eat straw like the ox." *(1)*

These descriptions refer to the condition which will prevail during the millennium reign of Christ when He will reign as King of Kings and Lord of Lords, and since that period of one thousand years returns the earth to the state it was in when it had been replenished by God, it is only but right that we can assume that these conditions did exist during the period of time prior to the fall of Adam.

We must now turn our thoughts to man who was created perfect and was given dominion over all the living things, which moved upon the face of the earth (Gen.1/28) and had, in fact, named all of them (Gen.2/20). In this dominion he was subject only to the divine will of the Almighty, and his dominion was **"...over all the earth..."** as is stated in Genesis 1/26.

This state of perfection is not to be measured by modern standards and, as it was the perfection which was created by the Almighty God, it can be said without compunction that Adam was perfect in the eyes of God. This perfection applied to every aspect of Adam's creation, namely, his physical, mental and spiritual standing in the sight of God. An example of this

Creation and Science Reconciled

is seen with his brain, which would have been in 100% condition and usage. It would undermine our understanding of the state of perfection which prevailed during this period, if we were to accept that Adam's brain power was less than 100%, in both condition and usage. Whereas modern medical science has indicated that the normal usage of the brain today is between the range of 10% to 20%. It is also accepted that there is no correlation between brain size and intelligence and it is stated that the brain of the Russian author Ivan Turgenov weighed 4.44lbs, whilst that of the French writer Anatole France weighed little more than half that figure, at 2.24lbs.

These facts must cause us to wonder at the abilities Adam had and used, and we must be careful not to encourage speculative thinking to influence our understanding of his ability. We may, however, be allowed to consider one example and that is today's evidence of telepathic powers being used and this raises the question, "Did Adam have the power and ability to transfer thought without speech?

After the 'fall' of Adam degeneration did set in, physically and spiritually, and an example of this can be seen in the Scripture in the loss of longevity. Reference should be made here to Genesis 5/3-32. That there are other examples of degeneration is without doubt, but these are the subjects of other studies.

There can be no doubt that man adapted as a result of climatic and dietary influences, resulting in differences in stature, skin colour and facial appearances amongst other physical characteristics, but these adaptations cannot be said to be evolutionary in any way.

We have looked at Adam's perfect creation and we should now look at a few of the other conditions, which prevailed at that time. The first recorded commandment given to Adam was in Genesis 1/28, and we must quote it here because of its importance. It reads:

"And God blessed them, and God said unto them; be fruitful and multiply and replenish the earth, and subdue it…"

We must come to the conclusion that this commandment was obeyed, with the result that the entire earth was fully populated by the offspring of Adam and the woman. That it was **'after their kind'** needs no further comment, and is the basis of our modern expression, in the use of the word "Mankind."

In looking at the offspring of Adam, we must accept that they were also in a state of perfection, as to think otherwise we would have to consider the situation of Man in a state of perfection producing an offspring which was not after his kind and, therefore, not perfect. This would have resulted in the untenable situation of imperfect beings co-habiting the earth, with the perfect man whom God had created. ***This is totally unacceptable.***

We can see that this state of perfection, both in Adam and in his offspring, was the root cause of the hatred which Satan had for them. Also, the perfection which existed throughout the replenished earth, caused Satan such displeasure that it was through the woman and the Serpent that his temptation of the woman brought about Adam's disobedience.

Let us now look at the man whom God had created.

Scripture tells us that God made man, not a youth or a child who would grow into maturity, but a fully mature man. It follows that the woman, who was formed out of the rib of the man, would be in full maturity also. These points are essential if we are to accept that Adam fulfilled the commandment given to him, to be, **'fruitful and to multiply and replenish the Earth.'**

Speculation could be made as to the age at which Adam was created, but to speculate would be wrong, in that, apart from our discussion above, there is no time scale which could be used to determine the age at which a man reaches maturity. This is accepted in modern thinking,, as being at or about the mid-thirties, but there are so many considerations to be made that even this age is open to question.

The acceptance of the full maturity of Adam at the time of his creation, leads us to the assurance that there was no ageing process during this period, apart from the maturing of an infant into a fully matured man or woman. This assurance can be borne out by the statement in Genesis 3/3, where we are told of the age of Adam when Seth was born and this is the first reference to any age being attributed to Adam. No age is given for when the births of Cain and Abel are recorded in Genesis 4/1 and 2 and we can therefore say, with real assurance, that the age of 130 given to Adam at the birth of Seth, began with the eviction of Adam and the woman from the Garden of Eden. This was the beginning of recorded Biblical time.

This is now an appropriate point in our study to point out that the woman was not known as Eve during this period and that this name was given to her by Adam, after they had disobeyed God. Prior to that event, they would both have been known by the name Adam, which was given to them by God, (Gen.5/2) which reads, **"Male and female created he them; and blessed them, and called their name Adam, in the day when they were created."**

This is the principle which remains to this day, in that when a couple is married, the name of the man is taken as their married name. Confirmation of the naming of Eve by Adam is given in Genesis 3/20 and it should be quoted here as it confirms that Adam and the woman did obey the commandment to, **"…be fruitful and multiply…"**.

The Scripture reads, **"And Adam called his wife's name Eve; because she was the mother of all living"**, and this does not, obviously refer to any creature, other than to Mankind.

This statement cannot be taken as being anything other than of an historic nature, as it refers to what was alive, of Mankind, at the time. It certainly does not have a prophetic overtone.

A point which can be accepted without being speculative or presumptive is that there was no illness or disease, whatsoever, on earth during this period. To say otherwise would only destroy our understanding of the perfection of the creative works done by God.

The obvious development of this thinking is that there was no occasion when death occurred during this period and, as none is recorded, we can assume with assurance and confidence, that none occurred. Consideration of this point, in comparison with what we know today, can only but cause us to reflect on the beauty of Creation before sin entered and brought about the downfall, not only of Adam and all mankind, but of the entire Creation which suffered as a result. We read of this latter point in Scripture and because of its importance, let us quote from Romans 8/22, which reads, **"For we know that the whole creation groaneth and travaileth in pain until now"**.

One thing which we can accept as being absent from the Garden of Eden, is that of fear and that absence of fear applied to the fish, the birds and the beasts as well as to man. *(2)* There was no need for one animal, or creature, to fear another, as they were not dependant on one another for food, as they were given: **"...every herb bearing seed...the fruit of a tree yielding seed; to you it shall be for meat"** (Gen.1.29).

That there was no fear of the Lord God by Adam, is to be taken from the time when Adam and the woman: **"...heard the voice of the Lord God walking in the garden in the cool of the day."** (Gen. 3/8)

It can also be taken that this walking in the garden was a normal occurrence, since the cause of the fear was not that of God but that through disobedience, they had eaten of the 'fruit of the tree of knowledge of good and evil' and this had shown them that they were naked. (Gen.3/10) This nakedness was a new experience for them and, verse 11 of Genesis 3, shows us the reasons for their knowledge.

This man Adam is known to us as the first Adam, and here let us quote again from the relevant Scripture, which reads, **"The first man Adam was made a living soul; the last Adam was made a quickening spirit."** (1st Corinthians 15/45)

Before drawing our study into this period to a close, we should look at the fact that whilst Adam had dominion over all the earth, his domain on earth was the Garden of Eden.

The 'act of faith' referred to in page 2 of the Preface allows the believer to build his/her life on the content of Scripture, knowing that it will not change as a result of any further revelation(s). This allows for the acceptance of the act of Creation without wondering how the Almighty actually carried out these works and also permits this type of study being carried out without questioning the omnipotence of God. Any questions as to ways of life; moral, ethical and spiritual can be answered by searching the Scriptures.

Evolutionary thinking, on the other hand, does depend on the information continually being discovered to be assimilated, or already formulated theories to be changed. That evolutionary discoveries can affect these established theories is a well-known fact, and this can lead to uncertainty developing in the minds of those who believe in this theory.

We should now examine a few of the various types of man, which have resulted from anthropological assessments of the finds of archeological excavations.

1. First homonid. The skull of this creature was found in Chad, Central Africa, and is about 5 million years old. It was announced on 10th July 2002 to be man's oldest ancestor and nicknamed Toumai (hope of life). It has features that are part ape and part human, but some experts have dismissed the discovery saying that the skull is that of a gorilla.

2. Homo erectus. A near complete skeleton of this was unearthed in 1984, and revealed a tall, big-brained smooth skinned homonid. It was believed to be about 1,800,000 years old, and to have come from Africa.

3. Archaic Homo sapiens. This type apparently emerged some 200,000 years ago in Africa, and is claimed to be a direct ancestor of our 'own species',

4. Homo sapiens. (Wise man) This apparently 'evolved' in Africa 130,000 years ago, and began colonising the rest of the world some 30,000 years later, reaching the other parts of the world between 60,000 and 40,000 years ago.

All of the comments relating to the separate types are based on the assumptions made by anthropologists on the finds made in the various digs and it can be seen that there is no finality of thought to them. One point common to each of the above is that it is said that they were from Africa, but the significance of this is not apparent.

They do compare with our findings in this study on one point only and that is they confirm that Man has been on the earth for much longer than the previously accepted time scale of 6,000 years.

With regard to the various digs being carried out, and there are apparently ten of these in Britain, one of the most important is that at Boxgrove in eastern England. Its importance is due to the number of *in situ* discoveries including, amongst them, the oldest human remains (a tibia and two teeth) found in Europe. Tools have been found in many other sites, but no conclusive human remains have yet been found.

This is a general synopsis of the grounds upon which the evolutionary theories are built and it is acknowledged that it is the priveledge, and responsibility, of each individual to decide to which 'camp' of knowledge or belief he/she subscribes.

It is of some importance to record here that there have been hoaxes in the field of evolution, which, if not discovered to be false, could have had a serious impact on the development of evolutionary thinking.

The two most documented are:

1. **The Piltdown Man.** This was the finding of a skull in England in 1912, but was only exposed as a hoax in the 1950s. This skull had purported to show that man's ancestors had man-like skulls and ape-like jaws, and was exposed because of finds in other parts of the world. These finds allowed for the dismissal of the claim that it had belonged to an early homonid. The main find was of that by Professor Raymond Dart of Witwatersand University, of a very different looking skull at the Taung caves in South Africa. Other bones from the species of this skull were discovered and doubts began to grow about the Piltdown Man. The species discovered by Pro-

fessor Dart is believed to have lived some 5,000,000 years ago.

2. **The Archaeoropter Hoax.** The author watched the television documentary regarding this hoax which concerned the manipulation of fossils and fabrication of evidence, to try to prove that the dinosaur had evolved into a bird. The manipulation, and reconstruction of the fossils was so expertly carried out that it fooled specialists in this field of science, so much so, that a magazine of excellent worldwide repute published an article about the archaeoropter. After exposure as a hoax this article was withdrawn, but only after considerable expense and time had been spent and wasted on the examination of the original evidence. The scientific value of this evidence was considerably damaged, although some further research was still able to be carried out on the manipulated fossils. The final analysis of this hoax was that there was no fossil evidence to show that a dinosaur had evolved into a bird.

If the evolutionary theory is worthy of belief, why is it necessary for such hoaxes to be perpetrated?

Before moving on from the subject of hoaxes, we must consider the fact that hoaxes were perpetrated in the field of religion.

In the main these were done to falsely prove the authenticity of relics of the religion, rather than the religion itself, and the most obvious of these is the 'Turin Shroud'.

This was supposed to have been the shroud in which the body of Christ was wrapped, (John 19/40) when buried in the tomb of Joseph of Arimathaea. It was later shown to have been a hoax, but this fact did neither nullify nor justify any of the beliefs in the Christian religion.

It should not be assumed that the only part of scientific thinking, which can be reconciled to Creation, is that of evolution. There are things such as the time scale of the earth's existence, fossils of creatures from before the 'Replenishment of the Earth (as discussed in Chapter 4) and other matters of real interest to believers and non-believers alike.

Let us now return to the discussion regarding Adam and the Garden of Eden and as we have the geographical boundaries of this garden described for us in Genesis 2/11-15, we should note that its boundaries were of rivers, which were the Pison, Gihon, Hiddekel and Euphrates. These rivers can be identified today (apart from the Pison) as being the following:

Gihon, which encompassed Ethiopia, (Gen. 2/12) is the River Nile.

Hiddekel, which was around Assyria, (Gen.2/14) is the Tigris.

Euphrates, which is still of that name (Gen.2/15) and, rising in the mountains of Ararat, flows down into the Persian Gulf.

Together with these river boundaries we can, by looking at a map of the Middle East, see that this land was bounded on the West by the Mediterranean Sea, on the South by the Red Sea and on the East and South by the Persian Gulf and the Indian Ocean. This land area can be compared with some real interest to the 'land' promised to Abraham.

It would be useful here to note that we are given no genealogy of Adam or of his descendants anywhere in Scripture relating to this period.

We will close this Chapter with the certain knowledge, that we cannot place any length of time to the events which we have discussed. There is no possibility of ascertaining the length of time which it took Adam to: **"Be fruitful and multi-**

ply and replenish the earth" but there can be no doubt that he did obey this; the first commandment which he received from the Almighty God.

APPENDIX TO CHAPTER 5

(1) Page 50. There was to be no carnivores in the creation?

The answer to this is an absolute *yes*, as we are told that Man and the other created beings were to be fed off the fruits of the herbs, trees etc. There were no carnivores in the creation.

(2) Page 54. Adam and Eve were without fear? It was slightly unfair to have created them this way, and then expect them to follow rules. After all, they would not understand or fear the consequences!

I do not see it this way as there was nothing in the Garden for them to fear. Adam was to be in dominion over the beasts of the field, the fowl of the air, and the fish of the sea. As they were not to be eaten for food, these creatures had no need of fear of man. Fear is a feeling of a purely negative nature and that is because it is generally caused by the unknown and, since man was given purely positive commandments, there was no need for him to be afraid.

His fear came about when he ate of the 'tree of knowledge of good and evil'. The commandment not to eat of this tree was positive, in that it said what not to **do**. When he ate of this, it gave him the knowledge that he and his wife were naked. This was completely new to him and he knew fear for the first time.

INDEPENDENT QUESTIONS RAISED BY THE WRITER OF THE CRITIQUE

(a) Did God bring about the destruction of the Original Creation that led to the 'void and dark' phase, or did Satan do it?

Satan's actions were the reason for God's decision to bring about the 'dark and void' situation.

(b) The fishes survived the destruction! - But there are many species, for which there are no fossils as old as the coelacanth fossils, which mean that they 'evolved' later.

This is not necessarily so as many types of fish were invertebrates and those which were vertebrates could have been buried under sand or silt and may not yet have been discovered.

Is this not an argument for 'evolution'?

I fail to see how this could be construed as such an argument as, in the case of the coelacanth, it was assumed extinct until a fossil was discovered. This confirms the thinking behind the study that science is still developing a theoretical approach whilst Scripture is complete and final.

(c) Are humans capable of their own stupidity and badness, or does Satan lend a hand? If so Satan is a very busy being.

When Adam was created, he was in a perfect state, and knew no evil. When he disobeyed God, he learned of the knowledge of good and evil and had the power to do one or the other and to abstain from the alternative.

The answer to the question is that humans are capable of

their own stupidity or 'badness' but Satan can and does lend a hand. He is a very busy being, but not omnipresent, in that he can only be in one place at a time. He has a host of 'fallen' angels who can, and do, assist in the evil that man does.

Conversely, human beings who are believers have the Comforter of John 14/16 and 17, and his ministering angels ever present with them to assist them in the problems which every day life brings.

Summary

General

The purpose of this Study is, as the Title states, an attempt to reconcile Creation and Science. This Study is based on Scriptural references, for the study of the Creative Acts of God, and for the Scientific aspect, reference has been made to established facts. Where possible these have been accepted for what they are, namely proof of the existence of fish, birds and beasts in a time frame, which is outwith that normally accepted by Christian Believers.

Whether or not this has been successful, the reader will decide for him/herself.

The basis on which the Study was started, is the acceptance of the fact that the Almighty God would not have created something which was 'without form and void', which is the description given of the Earth in Genesis 1/2.

Genesis 1/1 states that 'In the beginning God created the heaven and the earth' and it is thought to be worthy of Study, and to inquire, as to why the earth was without form and void.

Chapter 1 deals with the Original Creation of Genesis 1/1, and explores the information given in other Scriptures about this creative act. It will be seen that the Earth has been created to be inhabited, and we have seen that this habitation did occur. It was shown that the habitation referred to other 'beings' and apart from the various creatures listed below, and this aspect is covered, in some detail in Chapter 2.

That there were fish in the sea, beasts of the earth and birds of the air, and fossils of each of these, are there to prove that they had existed. The dating of these fossils allows for Time to be discussed, and this is one area of thought, in which reconciliation can occur between Scripture and Science. One aspect of this is, that the previously held date of 4004BC, which had been accepted as being the date on which the Creative works of God had been completed, is incorrect and the revised date of 6274BC will be shown as being the correct date.

The assumption that it was the completion of the works of Creation is also shown to be incorrect, and this is discussed in some detail.

That the earth was inhabited is the main subject discussed in Chapter 2, but it can be seen here, that there are at least two elements of Creation and Science, which can be reconciled, and those are that of Time and the existence of 'prehistoric' fish, beasts and birds.

It is of real interest to note that Time, an intangible element of both Creation and Science, which governs our daily life hour by hour, is also used to measure the vast distances of space. The scientific establishment of Time as the 4th dimension combines two factors of Creation, namely Time and Light, and enables the vastness of space to be used to demonstrate the infinity of Eternity.

Chapter 2 gives the Scriptural evidence of the inhabitation of Earth by angelic beings, under the domination of 'Lucifer, son of the morning'. His appearance, depth of knowledge and habitation are discussed in considerable detail, and this habitation is shown from the Scripture, as been on Earth.

The actual form and appearance of these 'angels' cannot be demonstrated, as there are no Scriptural references for this. There is one assumption which can be made, and that is that when Satan appeared to the Woman, it was in a form of an upright being. This is confirmed by the statement by God, in Genesis 3/14 to the serpent that, from that time on, the serpent would crawl on its belly. Had it already been in that state, there would have been no reason for God to command this to happen.

The reason for the 'waste and void' description of the Earth is also explored, and it is concluded that a cataclysmic occurrence caused this description to be an accurate one. The Scriptural reasons for the cataclysm are examined in some detail.

Lucifer is re-named Satan and his position, and that of his host of fallen angels, and their present state is covered in some detail.

Chapter 3 deals with the condition of the earth during the period, which must have existed between verses one and two of Genesis Chapter 1.

The absence of all life on the surface of the earth is discussed, but that fish life remained in the waters is accepted. Reference is made here to various Scriptures, and also Chapter 4 of this Study, which accept the fossil evidence of the now extinct life of the various sea creatures.

The continued omnipotence of the Almighty God is discussed in some detail, during this period of the earth's existence.

Chapter 4 discusses the Replenishment of the Earth on a *'day to day'* basis.

Since no mention is made in Scripture of the replenishment of any other part of Creation/ Universe, we can safely assume that the cataclysm which did occur did so only on earth.

It is seen that the progressive works by God, from the First Day through to the Sixth Day, have a form of logical development, which could be seen to be of a scientific nature.

Light was followed by 'an expanse' brought about by the

raising of the cloud base from zero to a level, which allowed the light to be of significance.

Dry land then appeared, and the dormant life was 'brought forth', to re-establish the flaura which then received direct sunlight to encourage growth.

The waters (seas) were then commanded to 'bring forth the moving creatures that hath life'. We discussed the creation of mammal life forms, which unlike other sea creatures, required the air formed in expanse of the Second Day.

This was followed by the new creation of beasts of the earth, and fowl of the air.

The final act of replenishment was that of a 'being' but not of the kind which had previously inhabited the earth, (see Chapter 2). This final act was the Creation of Mankind, (Male and female, as in Genesis 1/27) which we discussed in Chapter 5.

Chapter 5 tells us of the Creation of Mankind, and includes a discussion on the first commandment given to the Male and Female to "... be fruitful and multiply and replenish the earth...". This replenishment was with human beings, and is discussed in some detail. The chapter is included by a comparison of the differential thinking between Believers in Creation and those who accept the Theory of Evolution.